AF575550

Coral Martincavage

Creating Young Nonfiction Readers

EZ Readers lets children delve into nonfiction at beginning reading levels. Young readers are introduced to new concepts, facts, ideas, and vocabulary.

Tips for Reading Nonfiction with Beginning Readers

Talk about Nonfiction
Begin by explaining that nonfiction books give us information that is true. The book will be organized around a specific topic or idea, and we may learn new facts through reading.

Look at the Parts
Most nonfiction books have helpful features. Our *EZ Readers* include a Contents page, an Index, and color photographs. Share the purpose of these features with your reader.

Contents
Located at the front of a book, the Contents displays a list of the big ideas within the book and where to find them.

Index
An Index is an alphabetical list of topics and the page numbers where they are found.

Photos/Charts
A lot of information can be found by "reading" the charts and photos found within nonfiction text. Help your reader learn more about the different ways information can be displayed.

With a little help and guidance about reading nonfiction, you can feel good about introducing a young reader to the world of *EZ Readers* nonfiction books.

Mitchell Lane
PUBLISHERS

mitchelllane.com

2001 SW 31st Avenue
Hallandale, FL 33009
www.mitchelllane.com

First Edition, 2021.

Author: Coral Martincavage
Designer: Ed Morgan
Editor: Morgan Brody

Names/credits:
Title: Pulleys / by Coral Martincavage
Description: Hallandale, FL :
Mitchell Lane Publishers, [2021]

Series: Discovering STEM
Library bound ISBN: 978-1-68020-583-1
eBook ISBN: 978-1-68020-584-8

EZ readers is an imprint of Mitchell Lane Publishers.

Photo credits: Freepik.com, Shutterstock, pp. 4-5 Photo by Christopher Burns on Unsplash, pp. 6-7 Photo by Brett Jordan on Unsplash, pp. 8-9 Shutterstock, pp. 16-17 Shutterstock,

CONTENTS

Words in **bold** can be found in the Glossary.

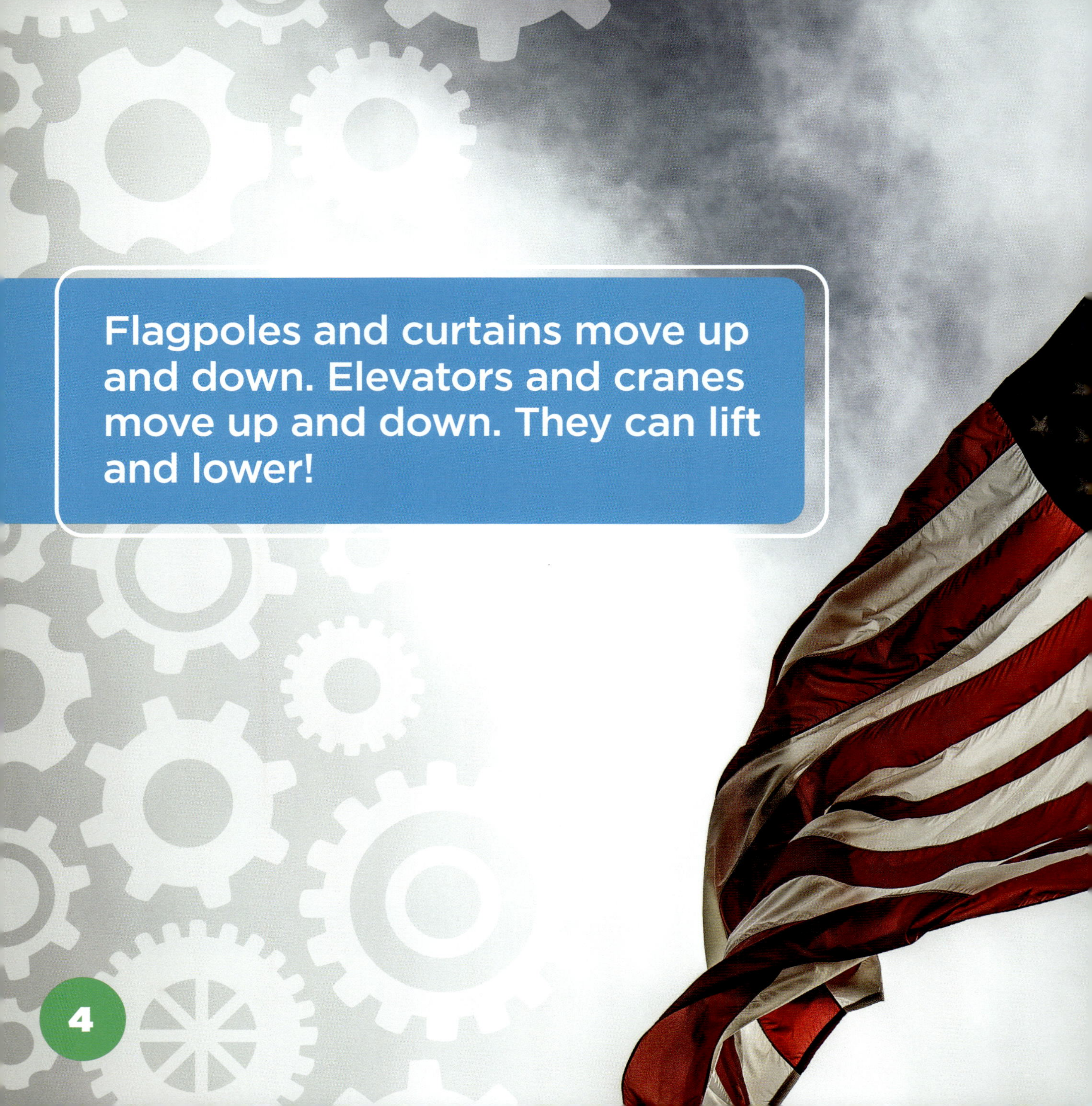

Flagpoles and curtains move up and down. Elevators and cranes move up and down. They can lift and lower!

They use **pulleys**. A pulley is a **simple machine**. Simple machines make work easier.

U.S.A.
HARKEN

A pulley is a wheel on an axle. Pulleys help **loads** move up and down. Pulleys move big loads with less **effort**.

The load is heavy. Pulling the rope is easier than lifting.

When I pull, the load moves up.
When I let go, the load drops.

Two pulleys can work together. Together, they move a bigger load.

Try using three. Now, use five! More pulleys make it easier.

The sails on a sailboat raise and lower. It uses a pulley.

Pulleys help us in many ways! You can find pulleys all around. What pulleys can you find?

INTERESTING FACTS

Pulleys were invented by Archimedes. He was a mathematician, engineer, inventor, **astronomer**, and **physicist**. This invention changed the way people moved objects. People in Mesopotamia used rope pulleys for hoisting water. In the 12th century, people used pulleys to build **cathedrals**. There are four main types of pulleys: Fixed, Movable, Compound, and Block and Tackle. Today, pulleys are used often in cranes, elevators, flagpoles, curtains, ladders, and exercise equipment.

GLOSSARY

astronomer
Person who studies the stars, planets, and other objects in space

cathedral
Church

effort
Physical energy needed to do something

load
The heavy object that is being lifted or moved

physicist
A scientist who studies matter and energy

pulley
A wheel with a rope on a grooved edge that is used to lift objects

simple machine
A device that makes work easier because it uses less effort to move an object

SOURCES

Pull, Lift, and Lower: A Book about Pulleys by Michael Dahl (Picture Window Books, 2006).

What are Pulleys? by Helen Frost (Capstone Press, 2001).

FURTHER READING

Use a Pulley! BY Mi-ar Lee (The ChoiceMaker Pty Limited, Big and Small Publishing, 2016).

What is a Pulley? by Lloyd Douglas (Children's Press, 2002).

INDEX

ABOUT THE AUTHOR

Coral Martincavage was born and raised in Miami, Florida. She was a teacher, literacy specialist, and curriculum developer before becoming an author. She is an avid reader and an advocate for STEM education.